Amazing Ecosystems

Copyright © by Harcourt, Inc.

All rights reserved. No part of this publication may be reproduced or transmitted in any form or by any means, electronic or mechanical, including photocopy, recording, or any information storage and retrieval system, without permission in writing from the publisher.

Requests for permission to make copies of any part of the work should be addressed to School Permissions and Copyrights, Harcourt, Inc., 6277 Sea Harbor Drive, Orlando, Florida 32887-6777. Fax: 407-345-2418.

HARCOURT and the Harcourt Logo are trademarks of Harcourt, Inc., registered in the United States of America and/or other jurisdictions.

Printed in Mexico

ISBN 978-0-15-362226-7

ISBN 0-15-362226-1

6 7 8 9 10   0908   16 15 14 13 12
4500356340

Harcourt
SCHOOL PUBLISHERS

Visit *The Learning Site!*
www.harcourtschool.com

Where We Live

Wherever you live, you are part of an **environment**. An environment is all the living and nonliving things that surround you. This includes the people in your family, your friends, and others who live and work near you. Animals and plants that live near you are part of your environment as well.

Nonliving things are also part of your environment. Water, air, soil, and weather are all examples of nonliving things that are around you.

Environments change all the time. Sometimes the changes are slow. When big changes in an environment's weather occur over a long period of time, certain plants and animals can no longer live there. Other changes can happen quickly, such as when a bridge or a house is built. Plants die and animals must quickly find a new home.

The animals and plants that once lived near this bridge have moved or died.

> These trees provide shade on a hot day. They also provide a home and food for animals.

Every living and nonliving thing in the environment is affected by other things in the environment. Think about a large tree near your home. You might remember a hot day when you stood under it to enjoy its shade. The same tree might also be a home for birds or insects.

Fast Fact

Did you ever realize how important grass is? It provides foods we need, such as rice, corn, and oats. It also is used to make sugar, bread, and even plastics!

But the tree needs something from the environment as well. It needs sun and water in order to live. In many ways, the things in every environment connect to other things.

MAIN IDEA AND DETAILS What things make up our environment?

Get Connected

Every environment has a system that connects living and nonliving things to one another. It is called an ecosystem. An **ecosystem** is a community and its physical environment.

Some ecosystems may be smaller than the tip of your finger. Other ecosystems may be bigger than the ocean. Ecosystems may have just a few living things or hundreds of thousands of living things.

Think of a crack in the sidewalk, a hole in a tree, or even the land under a bridge. You would find plants, insects, and other animals living there. You can see some of them easily, but others you can only see under a microscope.

> **Fast Fact**
>
> Bats can be a great help to people. A group of Mexican tail-free bats can eat up to 30,000 pounds of insects in just one night!

Bats sometimes live under bridges such as this.

A riverbank is home to insects and other animals that feed on the plants next to and under the water.

The nonliving things in an ecosystem might be air, soil, and water. In addition, the ecosystem has cold or warm temperatures depending on where it is. The ecosystem also has different amounts of rain and snow depending on where it is.

A large ecosystem, such as a river, has many living and nonliving things that affect one another. Fish and plants live under water. Plants, insects, and other animals live near the water. The river is affected by conditions, which include rainfall and temperature. Other nonliving things, such as rocks and soil, change as the river changes.

 MAIN IDEA AND DETAILS Give an example of an ecosystem.

Living Things

Each person in your home is an individual. The people in your home make up a **population**. A population is a group of the same kind of individuals living in the same ecosystem. A group of palm trees is a population. So is a group of silkworms.

Populations live among other populations. Swans might live on a pond with ducks. Because ducks are different from swans, they are a different population.

Some animals live in groups. People live in groups called families. Dogs living in the wild live in a group called a *pack*.

Some populations can live in only one place. Other populations, such as people, can live in many different ecosystems.

Different populations can live in the same ecosystem.

Every ecosystem has a **community**. A community is all the populations that live in the same place. A water community might have populations of insects, trees, fish, and other water animals.

The mangrove trees are homes for insects. The insects are food for fish. The roots of the trees are shelter for fish and other animals that live under water.

Communities show us how things depend on one another. Animals eat plants. They also help plants by adding nutrients to the soil where plants live. Plants, in turn, also provide shelter. Everything works together!

 MAIN IDEA AND DETAILS What is a population?

This mangrove tree helps living things above and below the water.

Change in Ecosystems

A good example of how animals depend on plants might be the last meal you ate. It probably had fruits or vegetables in it. Other animals, like birds, insects, and squirrels, depend on plants in another way. They make their homes in plants.

Plants also depend on animals. Animals that eat plants keep the plants from spreading too much. This leaves space for different plants to grow near one another.

Because animals and plants depend on each other, any change can have a big effect on an ecosystem. Some changes involve living things. Deer that live in a forest find food from plants and trees that grow there. If there are too many deer, there will not be enough plants and trees in the forest to provide food.

These deer left their usual environment because they had to find food.

This plant died because it could not get the water it needed.

Some deer will leave the forest to find other sources of food, and some will die from a lack of food. There will be fewer deer in the forest. When the plants and trees grow back, there will once again be enough food for the deer.

Some changes involve nonliving things. Have you ever lived through a dry summer with little rainfall? If so, you have seen grass or other plants die because they did not get enough water. Often the plants are home to animals. The animals might also die or move elsewhere for food and shelter.

 CAUSE AND EFFECT What would happen to the birds in your back yard if their food supply ran out?

Climate and Change

A climate is the pattern of weather in a place over a period of time. Your climate may be warm and sunny or cold and rainy. It depends on where you live.

The climate of a place affects its ecosystems. Some ecosystems provide homes to only a few living things. For example, caribou, a kind of reindeer, live in some of the coldest parts of the world. Most other animals need a warmer climate.

The climate of a tropical rain forest is warm and wet. Rain forests have the most diversity of all the ecosystems in the world. **Diversity** is the number of different kinds of living things in an ecosystem. In a tropical rain forest, you can find monkeys, birds, butterflies, flowers, and many other animals and plants.

Whether an environment has a lot of diversity or is made up of just a few living things, climate is important. Because climate includes rainfall, temperature, and sunlight, a change in any of these can have a serious effect.

Imagine if a desert changed from a dry climate to one with more rainfall. The plants and animals living there would die or move to an ecosystem where they could survive. After some time, other plants and animals that need more rainfall would take their place.

 CAUSE AND EFFECT How would a warm, wet climate affect the populations that could live in an ecosystem?

Tropical rain forests provide environments for a great diversity of living things.

For Better or Worse

Humans affect ecosystems. Unfortunately, sometimes what we do to improve our lives harms the living things around us. When we clear trees to build new homes or highways, we ruin the ecosystem of animals and plants.

Humans also cause pollution. **Pollution** happens when harmful substances mix with water, air, or soil. We use chemicals that ruin the soil and water. We use fuels in our cars and factories that harm the air.

Pollution is dangerous for us, too! It can make us ill. That is why we enact laws to repair some of the damage we have done to ecosystems. Many people recycle paper, metal, glass, and plastic. We can also use products for cleaning that do not harm the environment.

Once this area provided an environment for many living things.

Humans are trying to repair the environment. Many people work on **habitat restoration**, returning the natural environment to its original condition. Some people plant new trees to replace trees that have been cut down. Others work to clean up wetlands, so that the homes of plants and animals will be protected.

Focus Skill **COMPARE AND CONTRAST** **How would you compare the positive and negative ways in which humans affect ecosystems?**

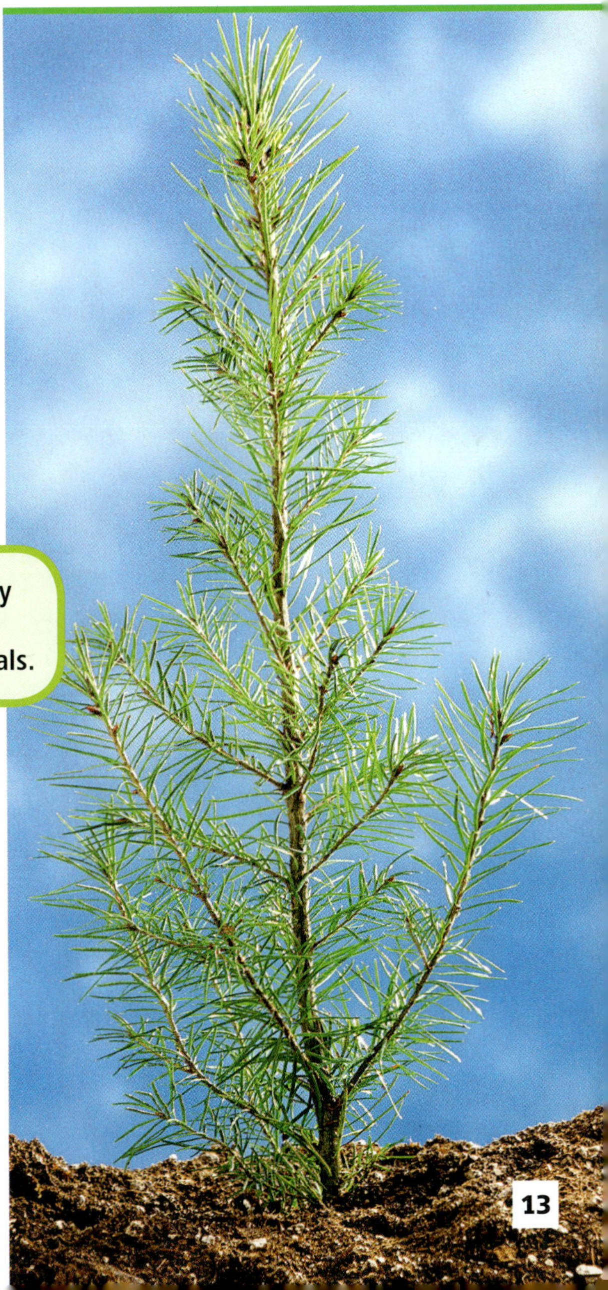

This tree will someday provide a home for birds and other animals.

Protecting the Air and Water

People have an impact on the environment around them. People can harm the environment in many ways. But they can also take steps to protect the environment.

Individuals and governments can both take steps to protect our air and water. Individuals can make a difference when they use resources responsibly. Governments can pass laws that make companies use resources in a responsible way.

You can help protect air and water by turning off the lights when you are not in a room. You can cut down on how much water you use. You can ride together with people and walk to destinations that are close by.

Governments can tell people not to pollute local waterways. They can pass laws so that factories will not pollute the air.

 COMPARE AND CONTRAST **How can governments and individuals help the environment?**

Summary

We live in environments that have ecosystems in which living and nonliving things are connected to one another. Changes in an ecosystem affect both animals and plants. Living things are adapted to changes in the environment. Humans have harmed ecosystems, but they can do things to restore and protect the environment.

Glossary

community (khu•MYOO•nuh•tee) All the populations of organisms living together in an environment (7)

diversity (duh•VER•suh•tee) A great variety of living things (10, 11)

ecosystem (EE•koh•sis•tuhm) A community and its physical environment (4, 5, 6, 7, 8, 10, 14, 15)

environment (en•VY•ruhn•muhnt) All of the living and nonliving things that affect an organism (2, 3, 4, 8, 10, 11, 13, 14, 15)

habitat restoration (HAB•i•tat res•tuh•RAY•shun) Returning a natural environment to its original condition (15)

pollution (puh•LOO•shuhn) Waste products that damage an ecosystem (14)

population (pahp•yuh•LAY•shuhn) All the individuals of the same kind living in the same environment (6, 7, 10)